BEHIND THE SCENES

LAUNCHING A SATELLITE

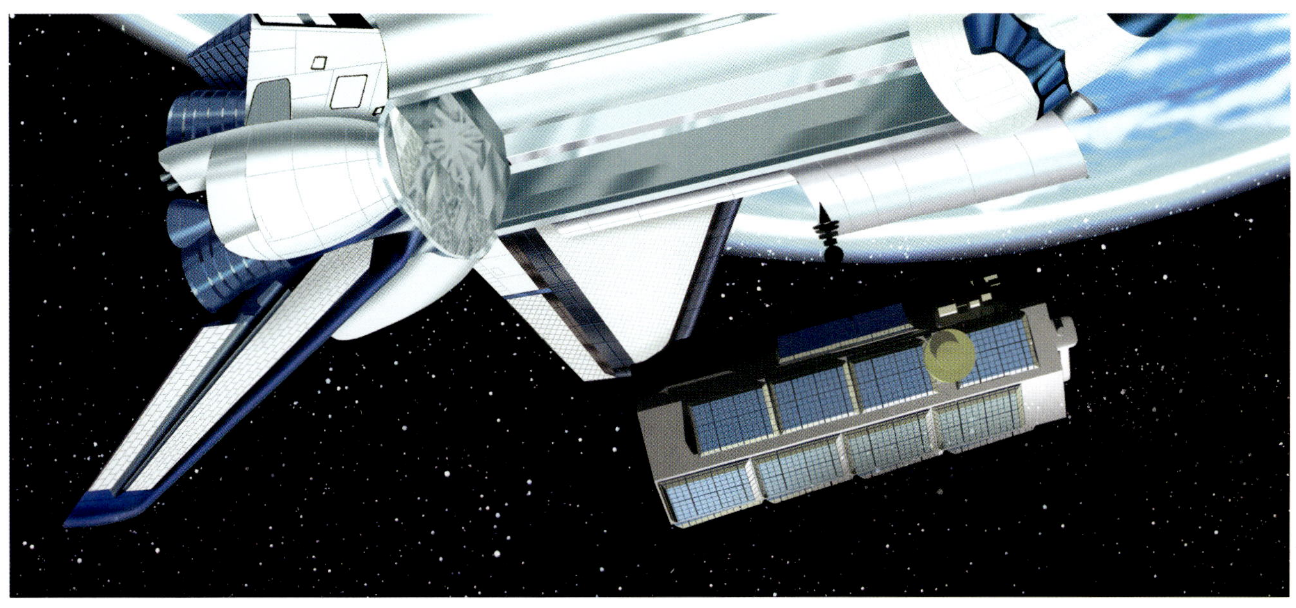

Written by Peter Mellett

Illustrated by Alex Pang

First published in Great Britain by Heinemann Library, Halley Court, Jordan Hill, Oxford OX2 8EJ, a division of Reed Educational and Professional Publishing Ltd.
Heinemann is a registered trademark of Reed Educational & Professional Publishing Limited.

OXFORD MELBOURNE AUCKLAND
IBADAN JOHANNESBURG BLANTYRE
GABORONE PORTSMOUTH NH (USA) CHICAGO

Produced for Heinemann Library by Lionheart Books, London.
Editor: Lionel Bender
Art Director: Ben White
Illustrated by: Alex Pang
Cover artwork: Roger Stewart
Page make-up: MW Graphics

Printed in Hong Kong by Wing King Tong

03 02 01 00 99
10 9 8 7 6 5 4 3 2 1

British Library Cataloguing in Publication Data
Mellett, Peter
Launching a Satellite. - (Behind the scenes)
1.Artificial satellites - Launching - Juvenile literature
I.Title
629.4'34

ISBN 0 431 02168 6

Acknowledgements
Every effort has been made to contact copyright holders of any material reproduced in this book. Any omissions will be rectified in subsequent printings if notice is given to the Publisher.

Any words appearing in the text in bold, **like this**, are explained in the Glossary.

Contents

Mission Control

A few weeks ago you entered a national school competition. You had to plan a scientific experiment that compares how things behave on Earth and in space. Hundreds of people applied – and your school won! Today is your first visit to the NASA Space Center in Florida, watching preparations for the latest flight of the Space Shuttle. The crew will launch one satellite and bring back another. They will also carry out some important experiments – and one of them is yours.

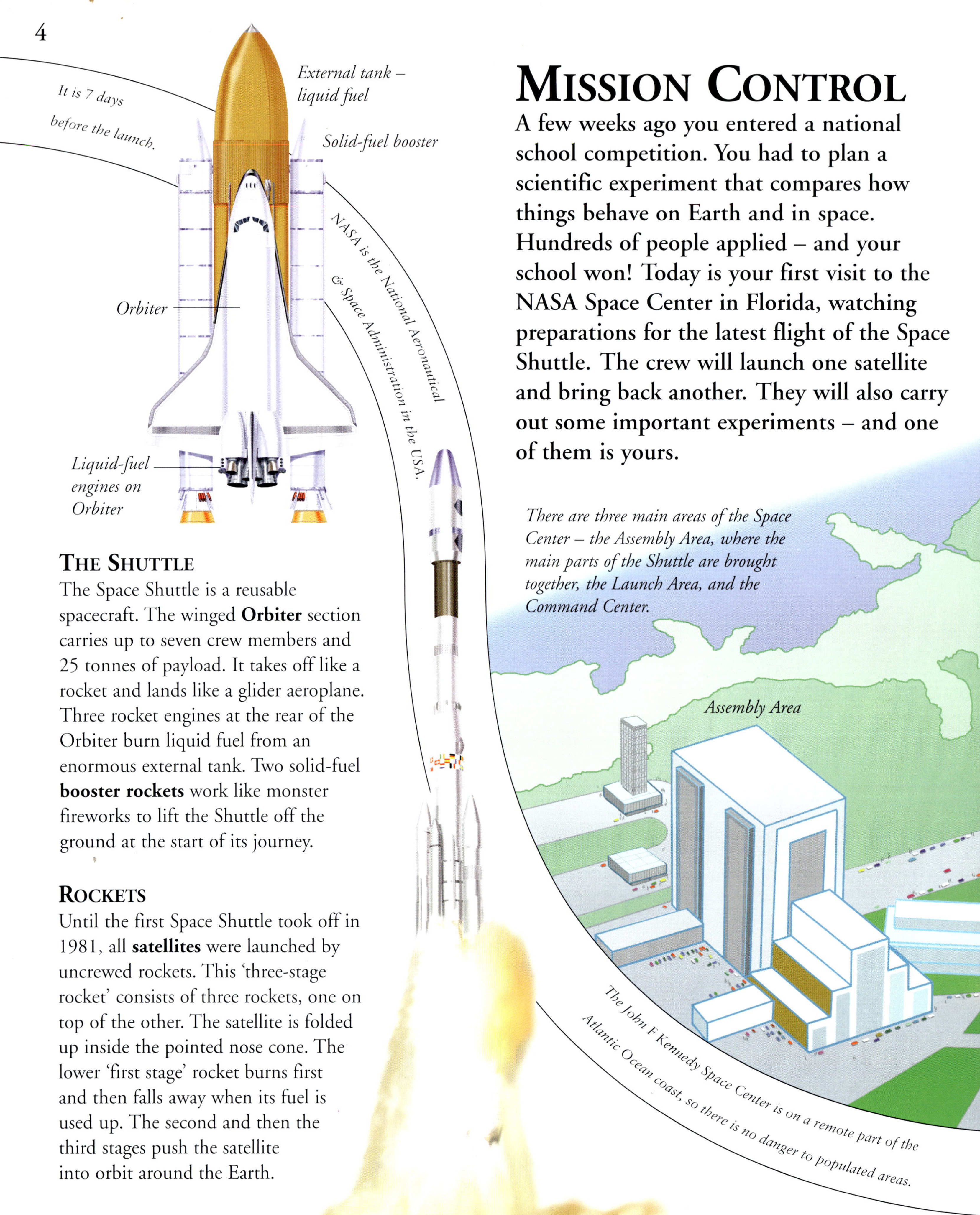

It is 7 days before the launch.

NASA is the National Aeronautical & Space Administration in the USA.

There are three main areas of the Space Center – the Assembly Area, where the main parts of the Shuttle are brought together, the Launch Area, and the Command Center.

The John F Kennedy Space Center is on a remote part of the Atlantic Ocean coast, so there is no danger to populated areas.

The Shuttle

The Space Shuttle is a reusable spacecraft. The winged **Orbiter** section carries up to seven crew members and 25 tonnes of payload. It takes off like a rocket and lands like a glider aeroplane. Three rocket engines at the rear of the Orbiter burn liquid fuel from an enormous external tank. Two solid-fuel **booster rockets** work like monster fireworks to lift the Shuttle off the ground at the start of its journey.

Rockets

Until the first Space Shuttle took off in 1981, all **satellites** were launched by uncrewed rockets. This 'three-stage rocket' consists of three rockets, one on top of the other. The satellite is folded up inside the pointed nose cone. The lower 'first stage' rocket burns first and then falls away when its fuel is used up. The second and then the third stages push the satellite into orbit around the Earth.

The satellite to launch

The crew's main task is to launch this satellite. It will give an early warning of **solar flares**. These flares are columns of white-hot gas leaping up from the Sun. They give off invisible **radiation** that often reaches the Earth. Unusually powerful solar flares can cause damage to satellites in space as well as cause power cuts and interfere with radio and TV signals.

NASA has formed close links with schools and colleges throughout the world.

The satellite to bring back

The Shuttle crew must also find a satellite and bring it back to Earth. It has been in orbit for five years, testing the effect of space on different materials. The satellite contains thin sheets of different plastics, metals and many other materials. It has also been carrying many different sorts of seeds. When the satellite is retrieved, schools across the country will receive seeds to test how well they grow.

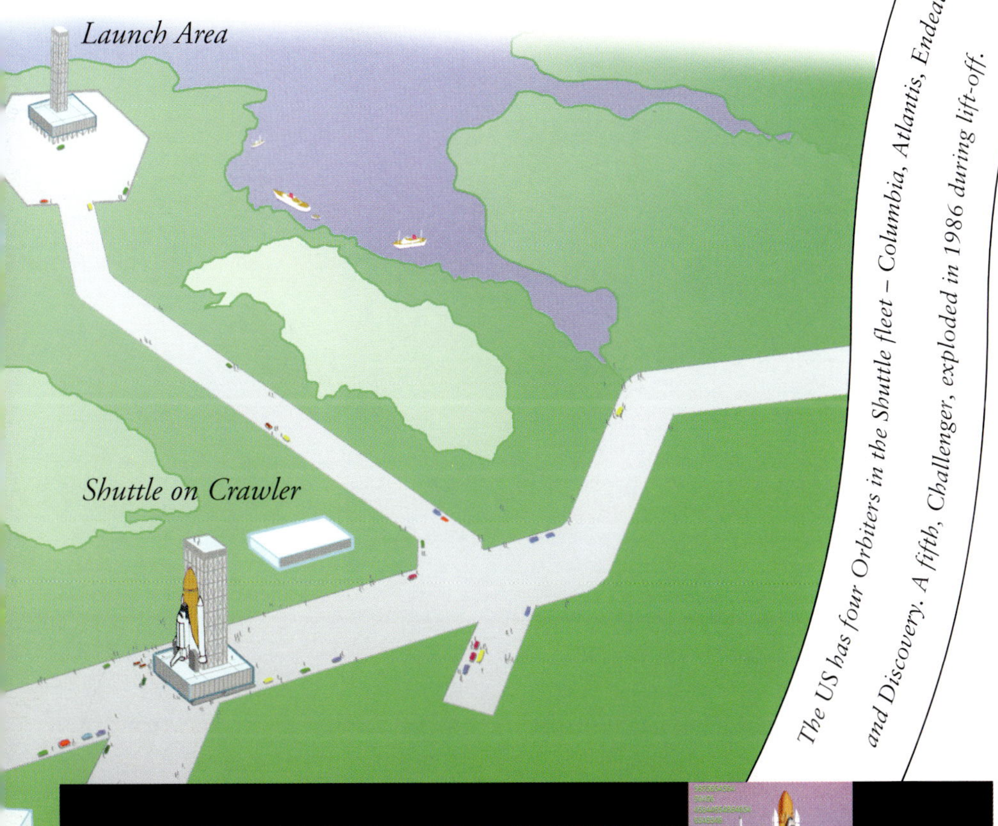

Launch Area

Shuttle on Crawler

The US has four Orbiters in the Shuttle fleet – Columbia, Atlantis, Endeavor and Discovery. A fifth, Challenger, exploded in 1986 during lift-off.

Your experiment

Here on Earth, helium-filled party balloons and the bubbles in lemonade float upwards. In space, there is no gravity and no direction you can call 'up'. How will balloons and bubbles behave there? TV cameras will watch the Shuttle crew in space and your friends at school carrying out the same experiment. At **Mission Control**, you will compare the results for a national TV science programme.

The Crew

Have you ever been on a roller-coaster and 'left your stomach behind' as you hurtle over a hump in the track? For an instant, you experience 'zero gravity' and become weightless. Orbiting the world on a curved path at 25,000 kph makes astronauts weightless all the time. These crew members are floating about inside an aircraft that is diving downwards like a roller-coaster. They have about two minutes of zero gravity to practise tasks they will perform later in inside the Space Shuttle.

Experiencing weightlessness

Astronauts in space suits practice handling a satellite underwater in a huge pool. With just the right amount of weight added to their suits, they neither float nor sink. This state is called 'neutral buoyancy' and gives some of the effects of **weightlessness** in space. Special underwater lighting makes the astronauts lose their sense of where 'up' and 'down' are, just like in space. An underwater expert makes sure the astronaut does not get into difficulties.

For this mission, the crew have practised for up to12 months.

The aircraft is an ordinary passenger plane with the seats removed and padding fitted to the walls.

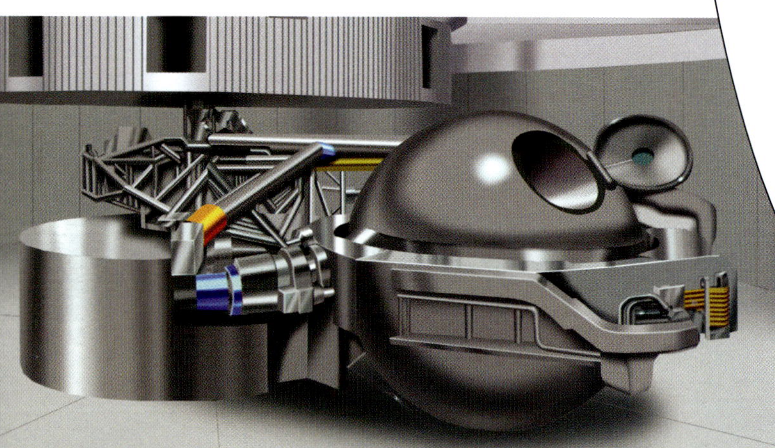

More people train for the mission than are needed. The final crew is listed just a few weeks before launch.

Experiencing take-off

Taking off in a rocket is like accelerating upwards in a lift – only the experience is much more extreme. When the Shuttle blasts off the ground, the weight of each crew member increases to three or four times the normal value. Whirling around inside this enormous spinning centrifuge gives the astronauts the same experience as taking off in the Shuttle.

Special tasks

Each crew member has been chosen because he or she has special skills that are needed on this mission. The **pilot** controls the Shuttle's flight in space and during landing. The **mission specialist** makes sure all the tasks are completed properly. The **mission commander** is in overall charge. The other crew members each have their own particular tasks to perform. All the crew members are tested to make sure they are especially fit and in excellent health.

This astronaut excercises on a treadmill while his heart-rate is carefully measured.

The pilot

The Shuttle pilot trained originally with the air force. He does not actually fly the Shuttle when it takes off from the ground. Everything is controlled automatically by computers. When the Shuttle returns from orbit, it flies like a huge glider. The pilot sits at the controls and guides it downwards.

The force on your body at the bottom of a 'vertical drop' roller coaster is greater than 4g.

Flight simulator

The mission commander and two crew members are practising inside a Shuttle simulator on the ground. As they alter the controls, the whole simulator moves as if it is flying. Even the view through the front windscreen changes. Everything the crew will do in space is practised time and time again on the ground.

The crew experience forces up to four times the force of normal gravity – called 4g – when the Shuttle lifts off.

Constructing and Testing

While the crew are practising for the mission, engineers are making the satellite ready for launching. Sensors at the front of the satellite will face the Sun. They will detect the radiation coming from solar flares. A type of computer called a microprocessor changes electric signals from the sensors into a special code. A radio transmitter sends the code to Earth with the help of the antennae and its reflector dishes. The antennae also receive messages from Earth that help to control the satellite.

Light in the night sky

Curtains of glowing coloured light sometimes appear in the night sky near the Earth's North and South Poles. The Sun gives off a steady stream of tiny particles that are charged with electricity. Solar flares increase the radiation and its strength. The particles are attracted towards the strongest parts of the Earth's **magnetic field** at the poles. The particles collide with the thin air 200 kilometres above the ground, causing a glow like the inside of a fluorescent lamp.

In the north, the night-time coloured lights are called the Northern Lights, or aurorae borealis; the Southern Lights are called aurorae australis.

A technician checks all the instrumentation before the launch.

Dish antennae are positioned on the top of the satellite. They are orientated to send and receive signals in every direction.

Dish antennae

THE SATELLITE

Panels with electronics and microprocessors

Solar panels change sunlight into electricity

Preparing the satellite

It has taken almost a year to build the satellite. Everything must be absolutely clean so that the complicated mechanisms are not affected by dust. The air in the workshop is filtered and the engineers wear clean disposable overalls. Solar panels on the finished satellite are folded up so the craft can fit inside its pod in the Shuttle cargo bay.

Shuttle 'turnaround' is 9 months; preparation for this mission is 3 years.

Checking the wiring

This engineer is designing the electronic circuits inside the satellite's **microprocessor**. The circuit diagram is a drawing that shows how all the separate parts are wired together. The microprocessor acts as the brain that controls the satellite. If one part fails, then another part switches on to take over its job. There are heavy metal shields around the microprocessor to protect it from the Sun's **radiation**.

Solar flares are most likely when black spots, called sunspots, appear on the Sun's surface. The number of sunspots reaches a maximum every 11 years.

Inside the test chamber

The satellite is being tested inside a special refrigerated chamber that copies the conditions of outer space. There is no air and the temperature is minus 250 degrees Celsius – six times colder than at the South Pole. Powerful lamps heat some parts of the satellite to represent the effects of the Sun. In another test, electric motors shake the satellite to make sure it will survive the vibrations and the acceleration it will experience at take-off.

Heat-protector tiles

During its return from space, **friction** from the rushing air heats the outside of the Orbiter. Some parts glow red-hot. Thousands of hard glassy **tiles** protect these parts and prevent damage to the metal skin underneath. Every tile must be checked before the next mission. The technician uses a special type of glue to stick on new replacement tiles. Each tile is specially curved to fit snugly onto its own particular place.

LOADING THE SHUTTLE

The Shuttle mission takes off today. Technicians have packed the satellite into the cargo bay of the winged Orbiter section. There are enough provisions of fuel, air, and food to last 10 days. The belly of the Orbiter is attached to the huge external liquid-fuel tank. On either side of the external tank are the solid-fuel booster rockets. The crew wear ordinary overalls instead of space suits because the Orbiter contains its own atmosphere like that of the Earth.

THE LAUNCH PAD

The whole Shuttle stands on the Launch Pad with its nose pointing straight up into the sky. The Shuttle was carried on a crawler from the Assembly Area to the launch site. The crawler is a flat platform mounted on wheels and pulled by a powerful tractor. At the Launch Pad, a tall tower connects fuel pipes and electric cables to the Shuttle. A lift inside the **launch tower** carries the crew to the front of the Orbiter that is now 35 metres above the ground.

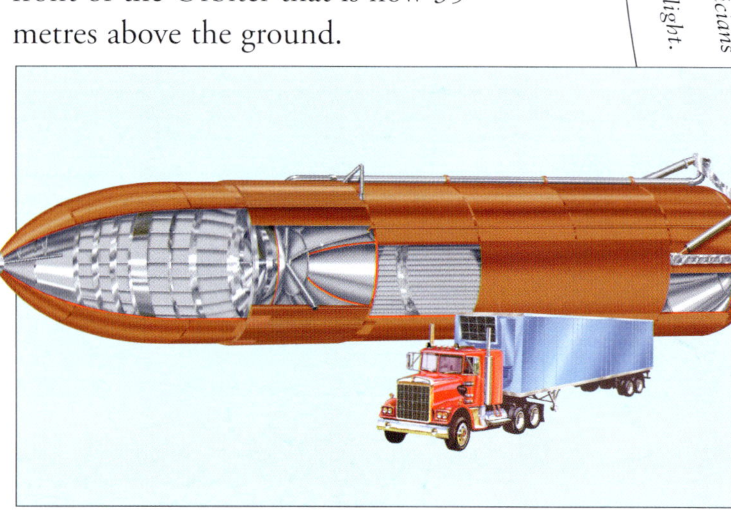

THE EXTERNAL FUEL TANK

The external fuel tank contains liquid **hydrogen** fuel. There is no air in space so the tank also contains liquid **oxygen** that helps the fuel to burn. The total volume of the two liquids is 2 million litres, about enough to fill an Olympic-size swimming pool. The temperature of these liquids is around 200 degrees below zero. The tank is 47 metres long and nearly 9 metres wide, equivalent to 20 large freight trucks stacked on top of each other.

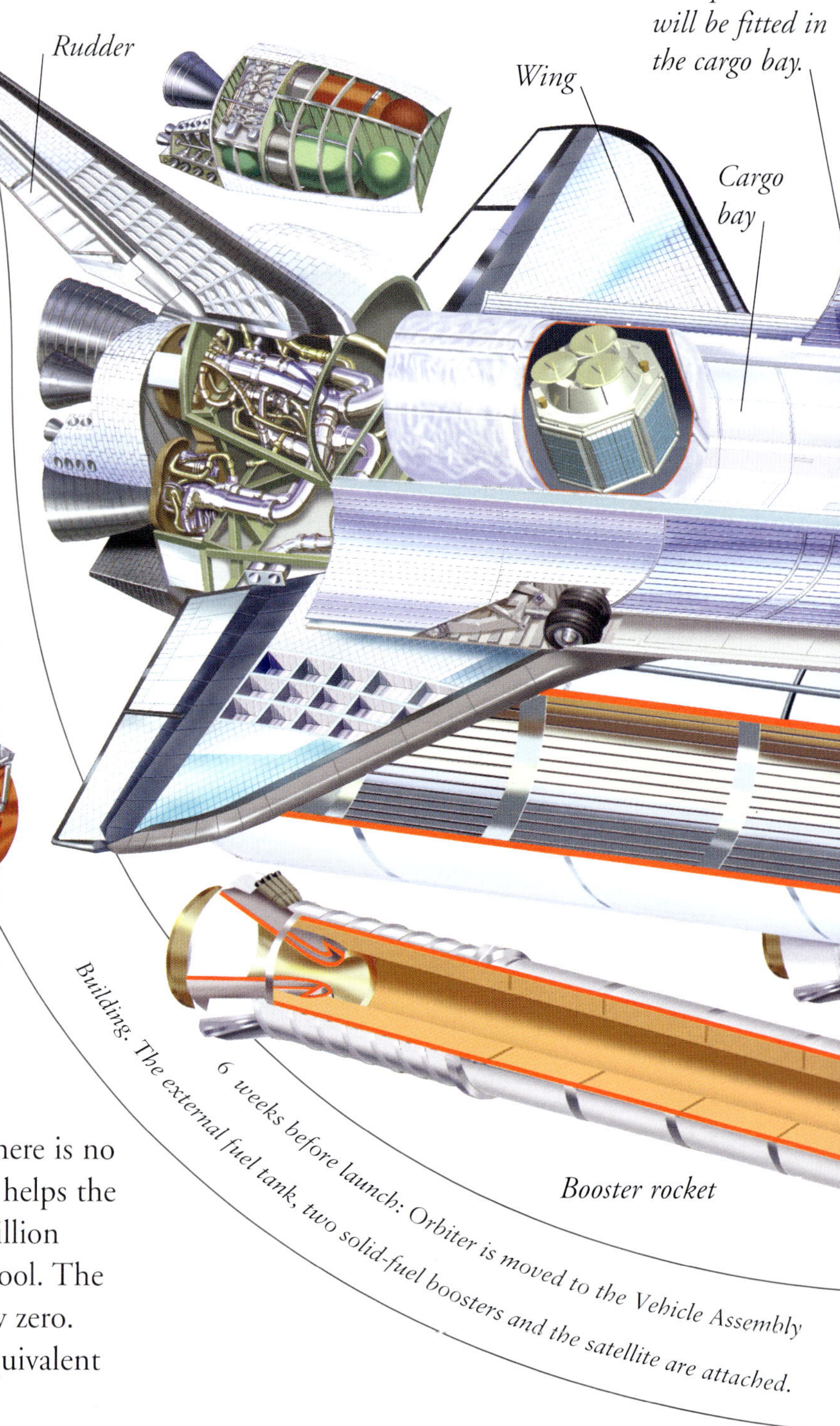

9 – 12 months before launch: technicians make the Shuttle Orbiter ready for its flight.

6 weeks before launch: Orbiter is moved to the Vehicle Assembly Building. The external fuel tank, two solid-fuel boosters and the satellite are attached.

Rocket engines

There are three liquid-fuel rocket engines attached to the tail of the Orbiter. Each of these engines uses a total mass of 500 kilograms (1600 litres) of hydrogen and oxygen every second (the same mass as a small car and the same volume as 11 bathfulls of water). The hydrogen burns with the oxygen to form water. Superhot steam roars out of the rocket exhaust, producing 250 tonnes of **thrust** – more than twice the power of a Jumbo Jet engine.

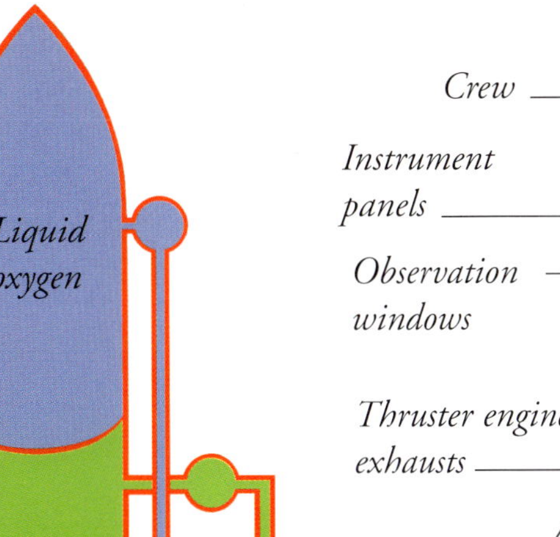

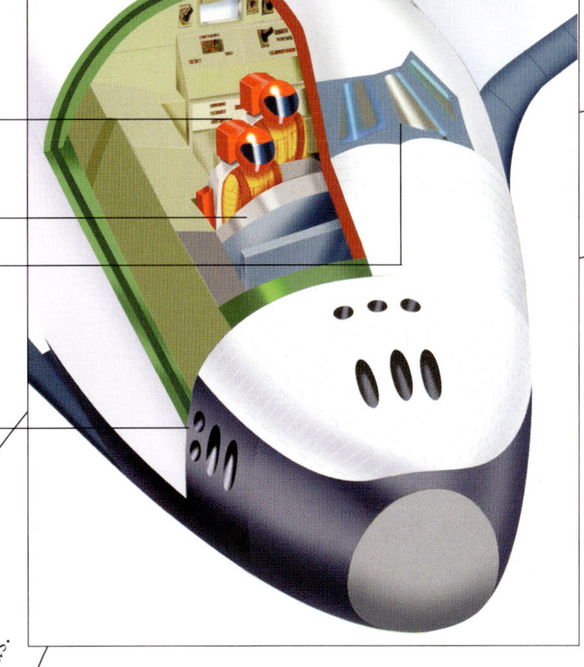

The flight deck

The crew settle into their seats more than an hour before take-off. Can you imagine sitting in a car or a passenger aircraft that is standing on its tail end? The Shuttle crew are lying on their backs with their legs pointing towards the sky. The hatch is sealed shut. Technicians at Mission Control help the crew to do hundreds of pre-flight checks. Take-off is in 10 minutes.

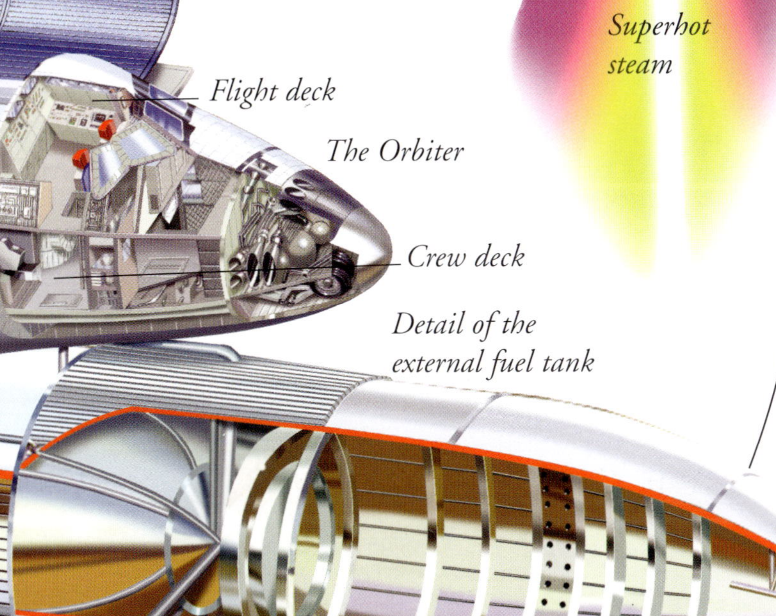

10 days before launch: the Shuttle is now in an upright position.

A giant tractor called a crawler carries it to one of the two launch pads.

Instrumentation

The **crew cabin** is at the front of the Orbiter. It has two main decks. The upper deck is the **flight deck** from where the crew control the Orbiter. The front of the flight deck looks like the cockpit of an airliner, with more than 2000 dials, switches, and press buttons. The crew live and sleep in the lower **crew deck**.

Take-off and Low Orbit

Everybody listens to the voice of the launch controller. "T minus 10 seconds and counting: 9 - 8 - 7 - ignition - 5 - 4 - 3 - 2 - 1 - we have lift-off!" The combined thrust of the five rocket engines is more than 3,000 tonnes. Gravity pulls down on the 2,000-tonne mass of the Shuttle. The upward force of the engines is greater than the downward pull of gravity so the Shuttle starts to climb away from the Launch Pad. After a few seconds, the Shuttle is accelerating four times faster than a racing car. After half a minute it is travelling faster than a rifle bullet – more than 2 kilometres a second.

Day 1 of the mission. The moment of launch is called T, so T minus 15 minutes is 15 minutes before launch.

Booster rockets

The fuel in the two booster rockets consists of aluminium powder moulded into a solid plastic cylinder. Mixed with the aluminium is a chemical that gives off oxygen when heated. The aluminium burns ferociously in the oxygen, vaporizing the plastic and sending a white stream of aluminium oxide particles and gases roaring from the exhaust nozzle. Have you ever held a powerful garden hose? The thrust of the rocket is in the opposite direction to the blast from the exhaust nozzle – just like the backward force of a water-jet shooting forwards.

In the last 6 seconds, the three liquid-fuel rockets start, followed by the two solid-fuel boosters.

THE EXTERNAL FUEL TANK

At 'T plus 2 minutes' the solid-fuel boosters are used up and they drop away. At 'T plus 8 minutes' the main engines shut down and the external fuel tank drops back down into the **atmosphere**. It tumbles and breaks up as it falls, parts splashing into the Indian Ocean 50 minutes later. The external fuel tank is the only part of the Shuttle that is not used again.

T+10 minutes – The Shuttle is 240 km above the Earth.

G-FORCE EFFECTS

The force from the rockets accelerates the Shuttle. Acceleration makes the Shuttle move faster and faster. The force, sometimes called '**g-force**', has an extra strange effect. Everything seems to weigh more. The crew members are forced back into their seats Even the skin on their faces is pulled out of shape.

RECYCLING THE ROCKETS

Each of the two solid-fuel booster rockets burns for two minutes. The spent rockets then fall away from the Shuttle while the liquid-fuelled engines continue to burn for another six minutes. Parachutes drop the used-up rockets gently into the sea. They are collected and packed with fresh solid fuel ready for the next Shuttle mission.

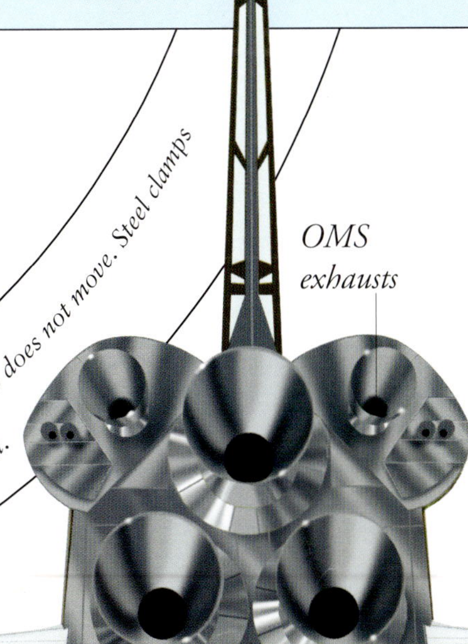

T minus 1 – the engines are producing full thrust but the Shuttle does not move. Steel clamps hold it down on the launch pad.

ORBITAL MANOEUVRING SYSTEM

The pilot fires the small rocket engines in the '**orbital manoeuvring system**' (OMS) in the tail of the Orbiter. When these engines shut down, the spacecraft is travelling at about 28,000 kph. This speed is called '**orbital velocity**' and is just fast enough to balance the Earth's gravity pulling downwards. The Shuttle now orbits the Earth continuously, without needing to use any fuel.

Launching The Satellite

The Orbiter is circling the Earth at a height of 250 km above the ground. All the engines are switched off and it orbits the Earth once every hour. It is now time to launch the satellite. The cargo bay doors have opened all along the back of the spacecraft. The satellite is inside a protective container. A crew member operates controls in the cockpit to open the container and release the satellite. The satellite has hardly started its journey into space. It has several million kilometres yet to travel.

Day 2 of the mission.

T+1 hour to T+20 hours: systems checks and crew acclimatization on reaching orbit: sleep and rest periods.

T+25 hours: opening cargo bay doors.

T+25 hours 30 minutes: opening satellite container.

A view from space of the Shuttle about to release the satellite. The Earth is seen below.

Releasing the satellite

A powerful spring pushes the satellite out from its protective container and away from the Orbiter. The satellite is spinning around like a **gyroscope** so that it does not wobble. When at a safe distance, the satellite rocket will fire, thrusting it into an orbit thousands of kilometres away from the Earth. At this distance it will have an uninterrupted view of the Sun.

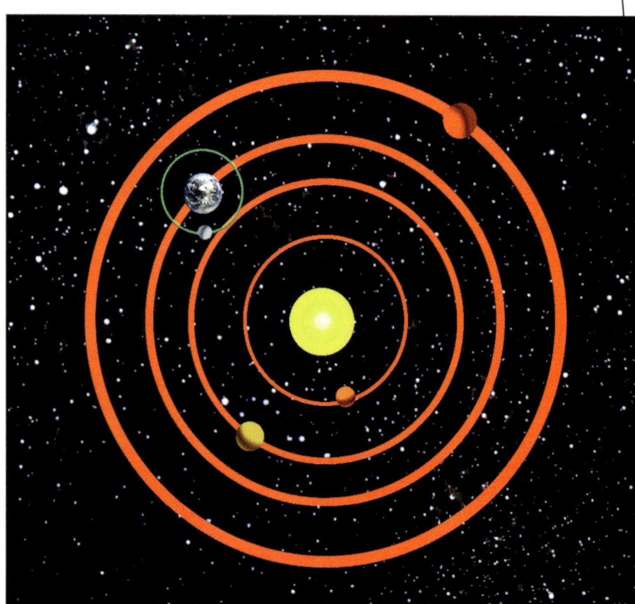

The inner planets

The Earth and all the other **planets** orbit around the Sun. Here you can see the four inner planets – Mercury closest to the Sun, then, in order, Venus, Earth, and Mars. The Moon is shown orbiting the Earth. There are five outer planets further out than the Earth. The satellite launched by the Shuttle will orbit closer to the Sun than the Earth does. When radiation sweeps out from a solar flare, it will reach the satellite up to a day before it arrives at the Earth.

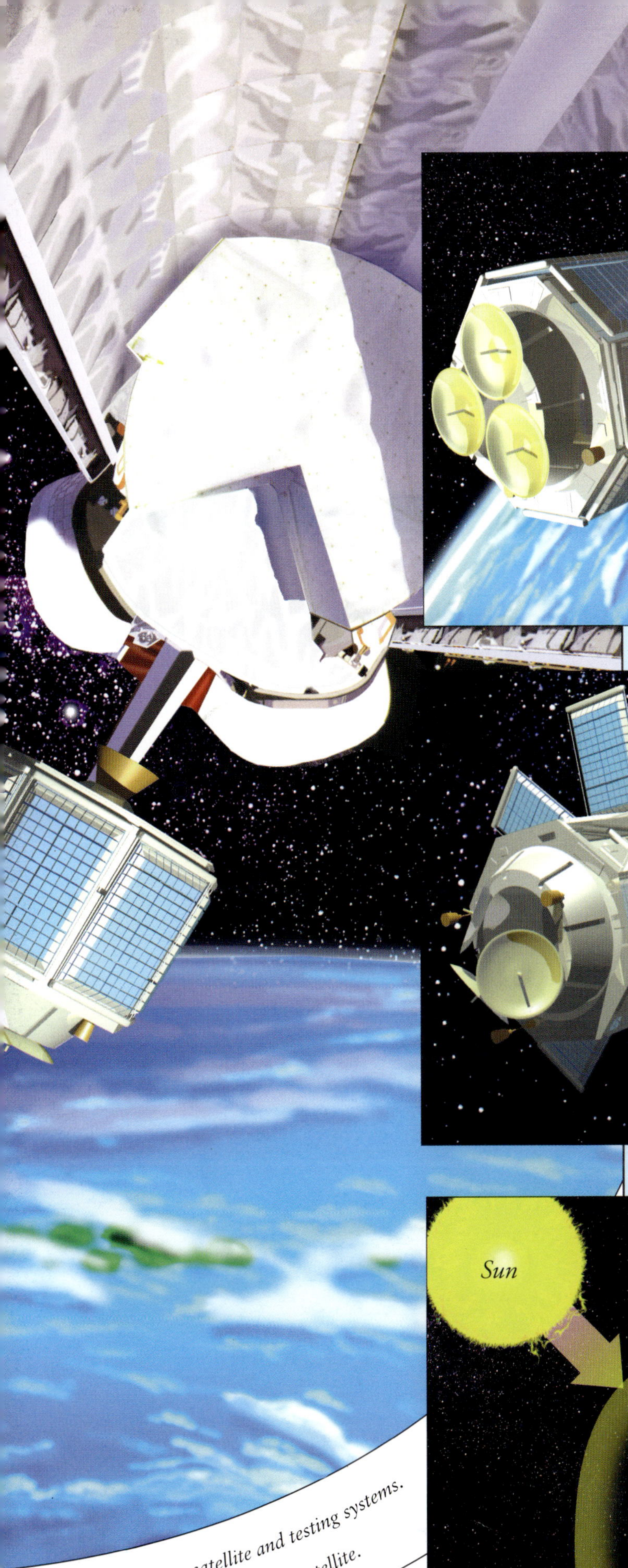

Satellite rocket

Launching the satellite from the ground would need an enormous rocket to overcome the force of gravity. From here, a very small rocket is powerful enough to send the satellite thousands of kilometres further away from Earth. The rocket burns for less than a minute and the satellite then coasts into its final orbit.

The satellite unfolds

The satellite was folded up for its journey so that the acceleration did not tear off fragile parts. When it reaches its planned orbit, the **reflector dishes** for the radio beams to Earth, automatically swing into place. There are **solar panels** fitted all around the outside of the **satellite**. They collect energy from sunlight and turn it into electricity to power the electronic equipment on board.

Waves of particles from the Sun

The satellite is fitted with detectors that sense radiation from solar flares. A camera gives scientists on Earth a clear view of the Sun. They look at the shapes of flares and predict if harmful radiation will come towards the Earth. Detectors on the satellite give advanced warning of high-speed particles called **ions**. The satellite **beams** a radio message to warn that the ions are coming.

Sun

wave of radiation

satellite

T+26 hours: ejecting satellite and testing systems.

T+27 hours 30 minutes: launching satellite.

Living in Space – Life-Support

Things do not fall in space. A pen hangs in the air where you leave it. Push gently and it travels at a steady speed and in a straight line until it hits the wall. The gravitational pull of the Earth is still there, but the speed of the Shuttle and every thing in it cancels out the downward force. If the Shuttle suddenly stopped dead on its circular path around the world, it would drop like a brick towards the surface of the Earth.

For long space flights, Shuttle scientists plan to recycle the astronauts' human waste and use it as compost to grow food crops.

In-flight food and drinks

Working, eating, and sleeping in space are done to a timetable worked out on Earth before the launch. There is not much room inside the crew cabin. Food is mostly semi-solid and paste-like so it does not scatter about. It is impossible to make a cup of coffee in a mug. Can you imagine what would happen to a spoonful of sugar? Drinks are sucked from collapsible squeezy packs.

Imagine the Earth is the size of a grapefruit. The Shuttle is 2 millimetres above its surface.

Gravity in an orbiting spacecraft is almost zero.

Neither up nor down

It can be fun to float effortlessly around inside the Orbiter, but weightlessness can be a problem. There is no sense of 'up' or 'down' and the result can often be a queasy, giddy feeling, like seasickness on a ship. Half of all astronauts suffer from '**space sickness**' but it passes after a few days when their bodies become used to the conditions in space.

It would be impossible to do this in space.

Sleeping bags

The Shuttle circles the world about 24 times each day. Night falls once every hour as the spacecraft swings into the shadow behind the Earth and the Sun's rays are blocked out. Thirty minutes later, the Shuttle emerges into the full light of the Sun. Like all human beings, crew members have an internal clock that is set to the rhythm of a 24-hour day. They sleep when their bodies tell them they are tired and that it is night-time back on Earth. They sleep in bags that are tethered to the walls.

Everything has the same mass as on Earth, but almost zero weight.

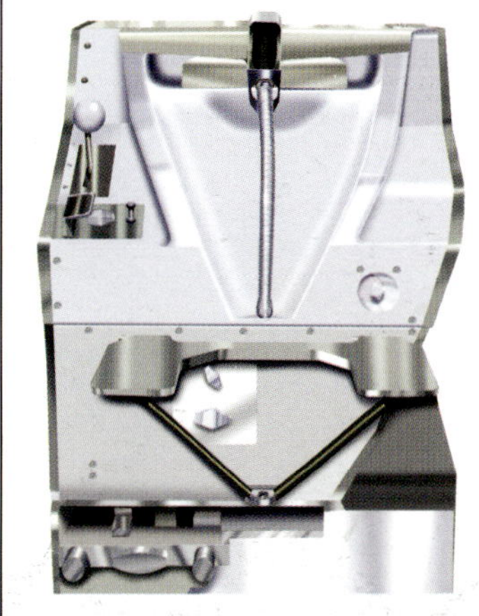

Front view of toilet

Air and water

The crew are supplied with air to breathe just like in the passenger cabin in an airliner. Filters remove the carbon dioxide the astronauts breath out. Special generators called **fuel cells** use **hydrogen** and **oxygen** gases to make electricity. Inside the fuel cells, chemical reactions combine these gases to make water that the astronauts can drink.

The washroom and toilet

Feeling hot and sticky? Need a wash? You cannot pour hot water into a basin in space. The cabin would fill with wobbling globules of water. All you can do is rub with a damp flannel. Lift the lid on the toilet bowl and a powerful fan starts. Any solid or liquid waste is instantly sucked down into the waterless pan. Filters separate solid waste that is dried and stored.

DANGER IN SPACE

The greatest dangers in space are the ones you cannot see. The Sun gives off rays called electromagnetic radiation. This radiation includes heat and light as well as invisible X-rays. High-energy X-rays from solar flares can pass through the outer skin of the Shuttle. Micrometeoroides are particles of dust that travel at immense speeds. On Earth, the atmosphere shields us from all these dangers. Micrometeoroids burn up as 'shooting stars' and the air absorbs most of the radiation. In space, astronauts must be specially protected.

The Sun's magnetic field – shown as the yellow lines of force and arrows – projects atomic particles into space – the purple arrows.

The Sun is made from hydrogen gas consisting of tiny hydrogen atoms made up from one proton and one electron.

SOLAR WIND

The Sun also gives off tiny particles called **electrons** and **protons** that carry charges of electricity. The Sun's magnetic field beams these particles out into space in a high-speed stream called the **solar wind**. The Earth moves through the solar wind as it orbits around the Sun. Billions of particles collide with the Shuttle every second as it orbits the Earth.

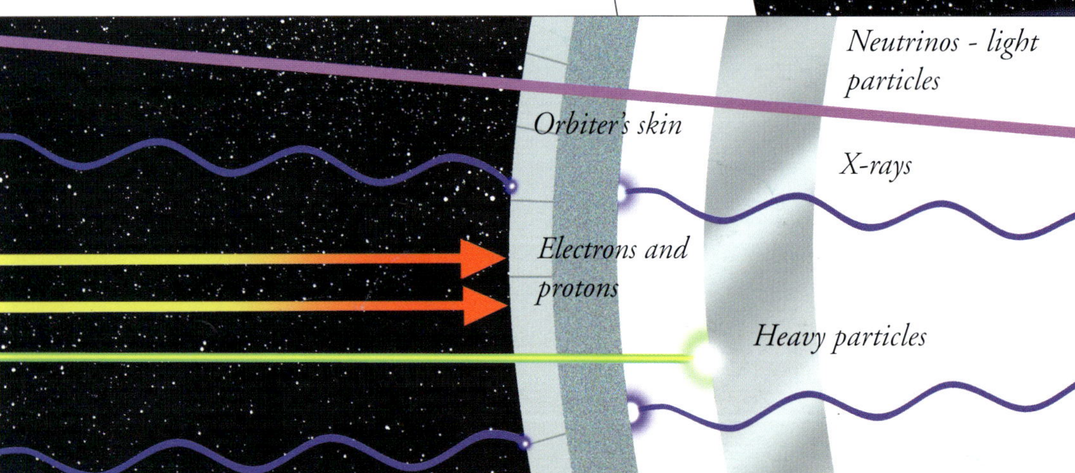

The immense temperature of the Sun makes the hydrogen into a soup of electrons and protons called plasma.

PROTECTIVE COVERING

The Orbiter's skin stops the heat as well as the electrons and protons normally radiated by the Sun. It cannot stop the heavier and more energetic particles from solar flares that can damage silicon chips inside computers. Some of these particles as well as **X-rays** can permanently damage the **living cells** in astronauts' bodies. Extra protection is needed if the mission lasts longer than a few days.

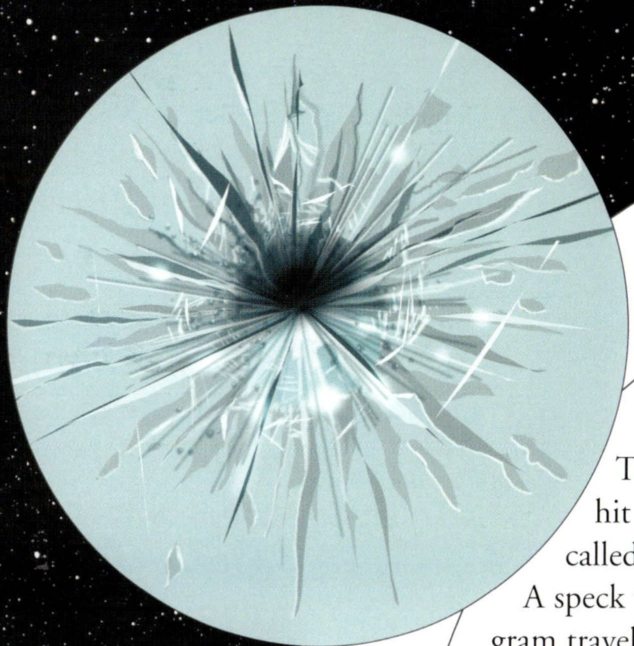

Damaged window

This Orbiter window was hit by a speck of space dust called a **micrometeoroid**.

A speck with mass of a fraction of a gram travelling at 10,000 kph packs the same punch as a 10 gram bullet travelling at 1,000 kph. The energy of a moving object depends on mass and speed – double the mass and you double the energy; double the speed and you quadruple the energy.

The Sun also throws out huge amounts of energy and showers of particles.

Hot and cold

The side of the Orbiter facing the Sun's rays is hotter than boiling water. The side facing the blackness of outer space is six times colder than the Earth's South Pole. The outer skin of hard **ceramic** tiles and an inner layer of insulating material protects the spacecraft against intense heat and cold. These layers also protect against micrometeoroids. The Shuttle is designed to survive a hole 13 millimetres across. Such a strike has never happened.

Locating space junk

The Earth is surrounded by millions of pieces of space junk, from dead satellites to tiny flecks of paint. Scientists pin-point the larger pieces by using **radar** dishes to bounce radio waves off the junk. They calculate the Shuttle's take-off direction and orbit to avoid them.

Navigating and Locating

Here is the satellite the crew will take back to Earth. How did they find it so easily in the vast emptiness of space? NASA has known its position since it was launched five years ago. Radio signals from the satellite help Mission Control to pinpoint its position. Computers work out how long the Orbiter's thrusters must burn and the direction of flight needed. Before the crew can see the satellite, radar signals from the Orbiter bounce off the satellite and show its position on a screen.

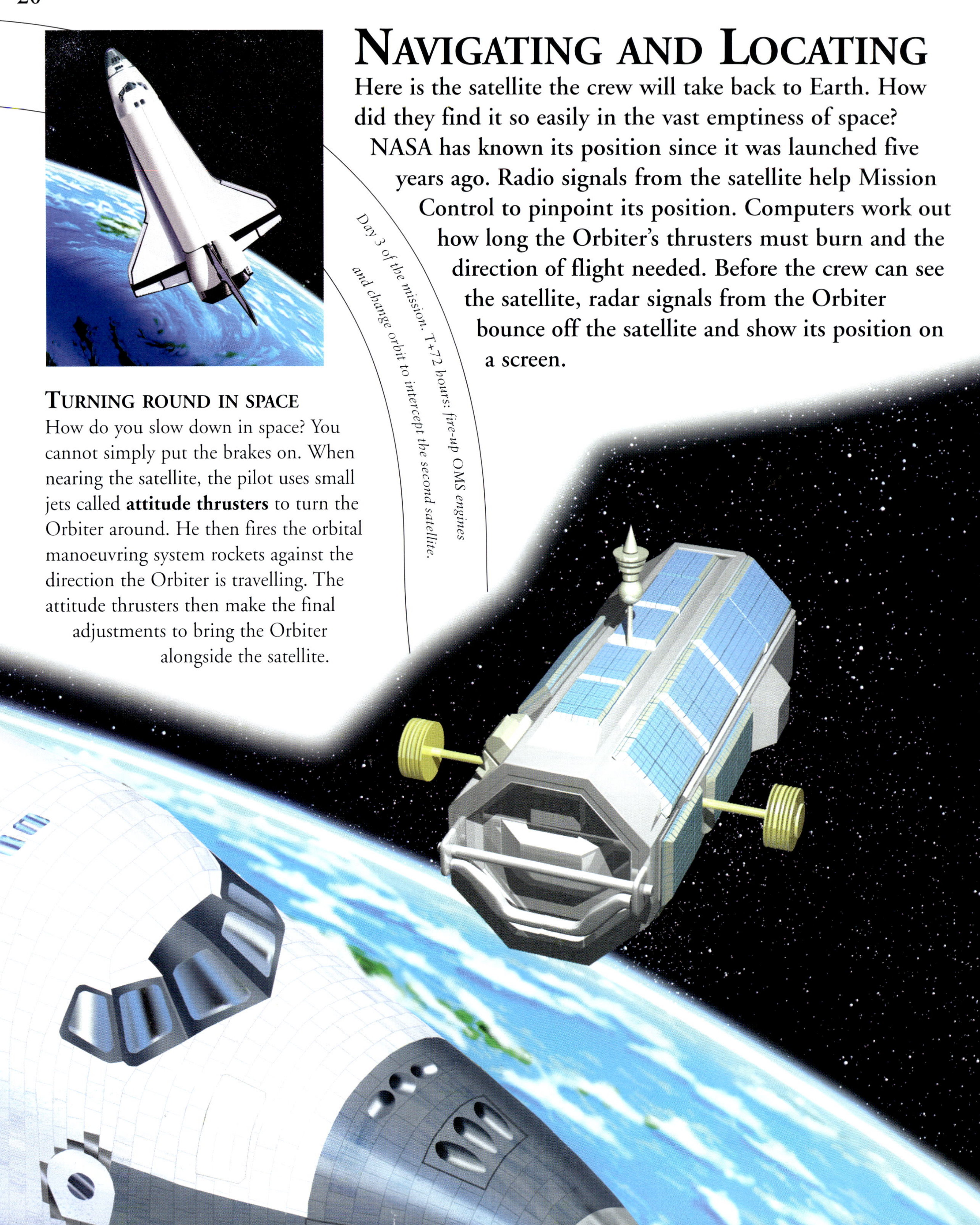

Day 3 of the mission. T+72 hours: fire-up OMS engines and change orbit to intercept the second satellite.

Turning round in space

How do you slow down in space? You cannot simply put the brakes on. When nearing the satellite, the pilot uses small jets called **attitude thrusters** to turn the Orbiter around. He then fires the orbital manoeuvring system rockets against the direction the Orbiter is travelling. The attitude thrusters then make the final adjustments to bring the Orbiter alongside the satellite.

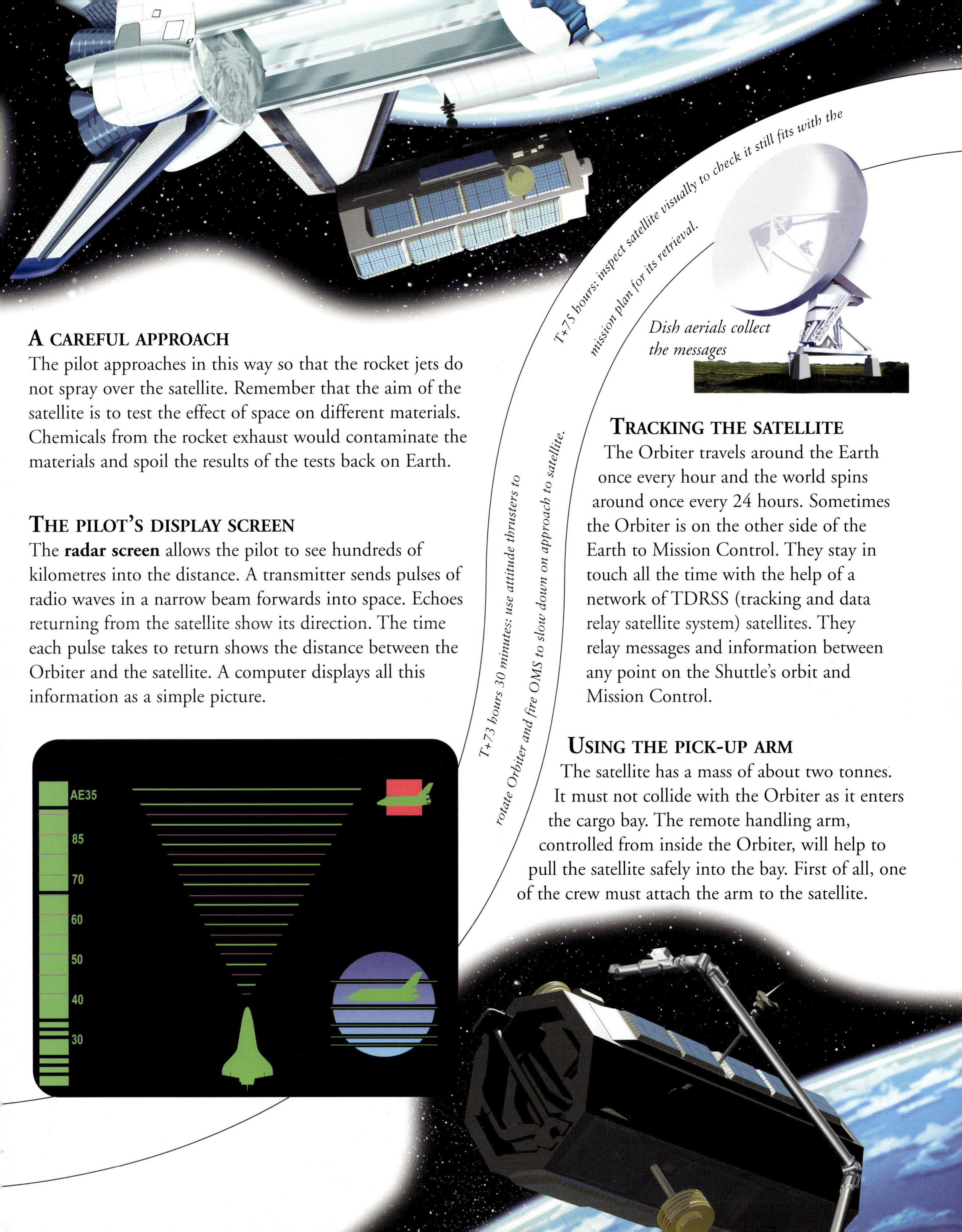

A careful approach

The pilot approaches in this way so that the rocket jets do not spray over the satellite. Remember that the aim of the satellite is to test the effect of space on different materials. Chemicals from the rocket exhaust would contaminate the materials and spoil the results of the tests back on Earth.

The pilot's display screen

The **radar screen** allows the pilot to see hundreds of kilometres into the distance. A transmitter sends pulses of radio waves in a narrow beam forwards into space. Echoes returning from the satellite show its direction. The time each pulse takes to return shows the distance between the Orbiter and the satellite. A computer displays all this information as a simple picture.

Tracking the satellite

The Orbiter travels around the Earth once every hour and the world spins around once every 24 hours. Sometimes the Orbiter is on the other side of the Earth to Mission Control. They stay in touch all the time with the help of a network of TDRSS (tracking and data relay satellite system) satellites. They relay messages and information between any point on the Shuttle's orbit and Mission Control.

Using the pick-up arm

The satellite has a mass of about two tonnes. It must not collide with the Orbiter as it enters the cargo bay. The remote handling arm, controlled from inside the Orbiter, will help to pull the satellite safely into the bay. First of all, one of the crew must attach the arm to the satellite.

Satellite Retrieval

The space suit backpack provides air to breathe for up to 6 hours and regulates the temperature inside the suit.

Small jets of gas from the backpack propel the astronaut towards the satellite. The space suit protects the astronaut outside the Orbiter from the extreme heat and cold of space. The astronaut must attach the end of the remote handling arm to the satellite. The Orbiter, the satellite, and the astronaut are all hurtling around the Earth at 25,000 kph. They seem stationary with respect to each other because they are all travelling in the same direction at the same speed.

Airlocks

The Orbiter is full of pressurized air for the crew to breathe. There is no air in space so the pressure there is zero. To leave the Orbiter, the astronaut must pass through an **airlock** that has an inner and an outer door. The airlock makes sure that very little air is lost from the Orbiter cabin. When leaving, the inner door closes before the outer door opens. When the astronaut returns, the outer door closes before the inner door opens.

Temperature control

A set of long underwear is worn next to the astronaut's skin. Small tubes are woven into the material. Temperature-controlled liquid flows through the tubes to warm or cool the astronaut's body. Remember that our bodies are at 37 degrees Celsius. Outside the Orbiter, the temperature in the shade is 250 degrees below zero, while direct sunlight is more than 400 degrees hotter.

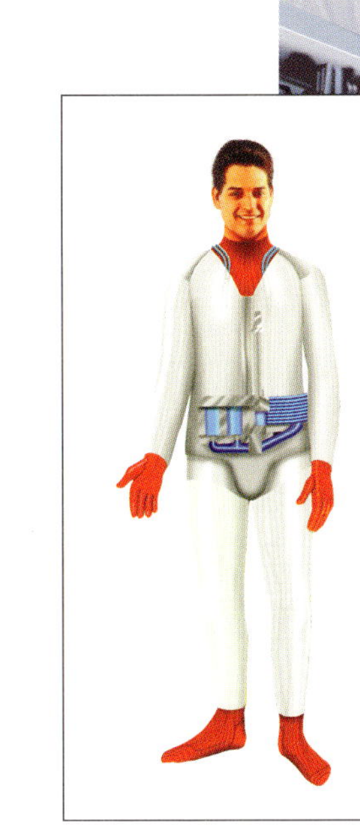

T+79 hours: satellite stowed in cargo bay

The satellite has a mass of more than 2 tonnes. It must be moved very slowly. If it hits the cargo bay doors, it will cause considerable damage.

The space suit

The main part of a space suit is the airtight pressure garment. It is made from stiff material coated with a type of rubber called neoprene. There are flexible joints to allow the astronaut to move. The third layer is like bullet-proof vest material and protects against micrometeoroids. On top are two layers of tough **Teflon** fabric. The space suit is technically known as an '**extravehicular mobility unit**' or EMU.

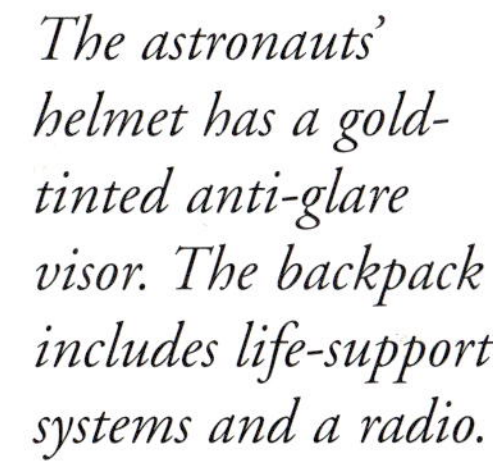

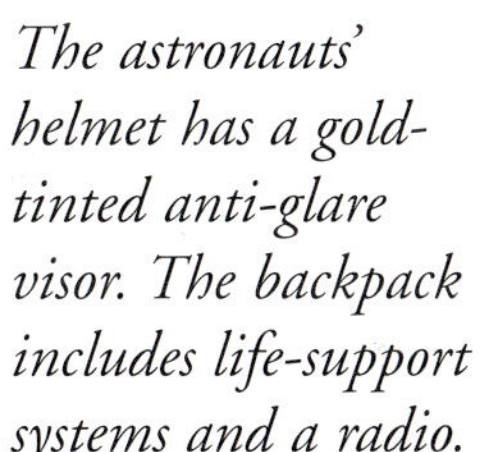

The astronauts' helmet has a gold-tinted anti-glare visor. The backpack includes life-support systems and a radio.

T+77 hours 45 minutes: astronaut leaves airlock and jets across to satellite.

Pulling in the satellite

The astronaut has attached the satellite to the end of the remote handling arm. Another crew member controls the arm from inside the main cabin and slowly pulls the satellite towards the Orbiter. The astronaut must carefully guide the satellite into its place in the cargo bay.

Using the backpack

For the second part of his mission, the astronaut practises flying up to 200 metres away from the Orbiter. The jet-powered backpack is a new design. Technicians at Mission Control want to make sure it performs properly in space. Astronauts on future missions may use jet backpacks to help assemble a space station.

T+76 hours: astronaut puts on space suit.

Carrying Out Experiments

Days 4 and 5 of the mission.

The crew have successfully launched one satellite and recovered another. It is now time to carry out the experiments planned earlier. The flight deck and living quarters at the front of the Orbiter are very cramped. Attached to the airlock in the cargo bay is a large cylinder-shaped compartment called 'Spacelab'. Here, the experiments are carried out. Some of the experiments are for commercial companies. The money helps NASA pay for the Shuttle missions.

Density is defined as: the mass of 1 metre cubed of a substance. The units are kilograms per cubic metre (kg/m³).

Spacelab fits in the cargo bay directly behind the flight deck.

Densities: aluminium = 2,700 kg/m³; iron = 7,860 kg/m³; helium = 167 kg/m³; water = 1,000 kg/m³.

Experimental equipment

Technicians set up the experiments inside **Spacelab** before the Shuttle was launched. Carefully packed inside the childrens' own labelled container is a bottle of lemonade, a helium-filled party balloon, and a piece of string.

Preparing special materials

Scientists have discovered that super-pure silicon crystals help to make super-fast computers. The best way of making very pure substances is to heat them in a special way in zero gravity. Most Shuttle missions spend some time preparing special materials in this way for use back on Earth.

On Earth, a balloon filled with helium gas rises upwards.

While the caps are screwed on the bottles, bubbles inside do not move.

Experiments in the classroom

The title of your school experiments was "Bubbles and balloons". Your idea came from watching party balloons and the bubbles in fizzy drinks - how do they behave in space where there is no gravity? Will there be any differences and can we explain them easily? You watch the pictures coming from space and your classroom and explain to the viewers what is happening.

A Shuttle made from iron could not carry enough fuel to reach orbital velocity. Iron is 3 times as dense as aluminium. Most of the Shuttle is made from aluminium: light and strong.

Watch the bubbles

Fizzy lemonade contains pressurized gas dissolved in flavoured water. Unscrew the cap and the pressure in the bottle falls. Bubbles form on the inside surface of the bottle. The bubbles are lighter (less dense) than the liquid. On Earth they float to the surface. In space they stay where they form, simply growing bigger.

Cargo bay door

Manufacturers' experiments and equipment

Access to flight deck

Experiment containers

The same experiments in space

Blow up a party balloon with helium gas and it floats away upwards. Helium gas is less dense than the air around. A balloon full of air has a mass eight times greater than the same balloon full of helium. The balloon floats upwards on Earth. In space, it hovers in front of the astronaut, just like her pen. There is no force of gravity inside the Orbiter. Everything is 'weightless', whatever its mass.

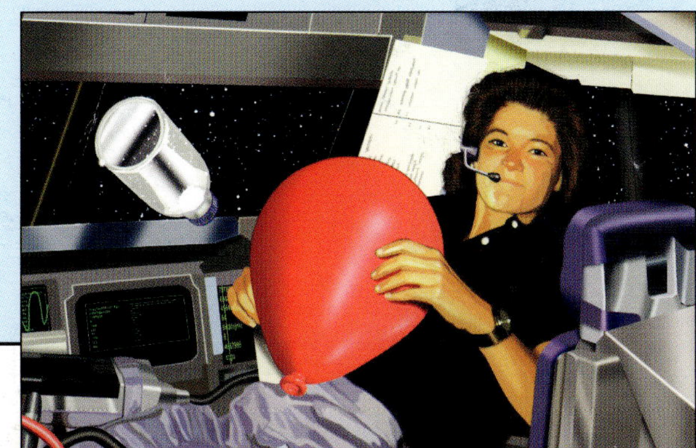

Leaving Orbit and Landing

After six days in orbit, it is time to return to Earth. The crew has successfully completed all the tasks planned for the flight. Up in space, a weather satellite looks down. It sends photographs to Earth that show freak storms heading towards the landing site. The Orbiter can make an emergency landing at another place but then it must be transported back to the launch site. If all goes according to plan, it will land back at the NASA Space Center just 12 hours before the storm is due.

As the Shuttle descends, deceleration forces can exceed 40g but rarely go above 10g.

The pilot turns the craft so that the wings skim the atmosphere.

Attitude thrusters

The Orbiter stays in space because it is speeding along a curved path around the Earth called an orbit. This high-speed movement has a stronger force on the Shuttle than gravity. In order to leave orbit, the commander uses the attitude thrusters to turn the Orbiter so that the OMS rockets point forwards. Firing the rockets slows the Orbiter and it drops towards the Earth. **Re-entry** has started.

The flighpath

Now 30 kilometres high and travelling twenty-nine times faster than a jet airliner (26,000 kph), the Orbiter dips into the thin upper atmosphere of the Earth. The air rubs against the Orbiter's skin. Friction heats the air and breaks it into particles called ions that have electric charges. The ions interfere with the radio equipment so that the Orbiter cannot communicate with Mission Control for several minutes.

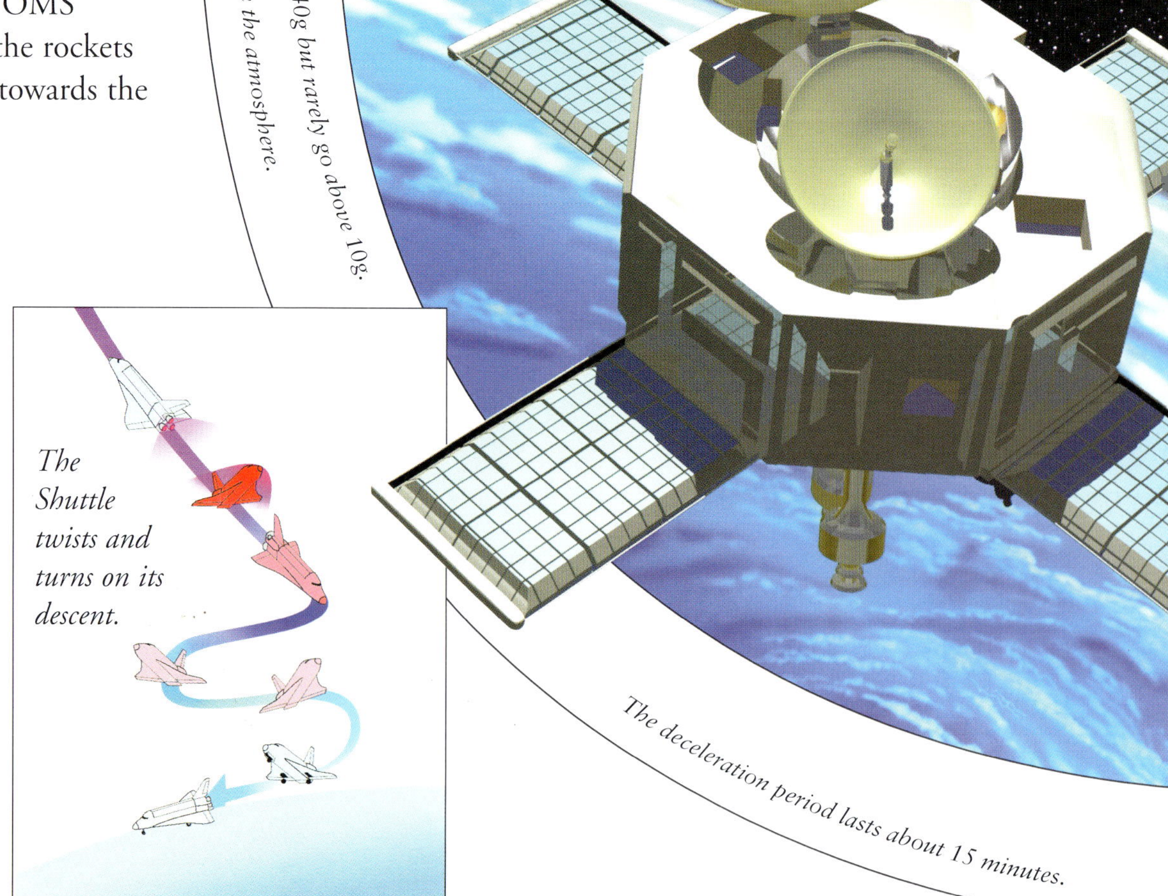

The Shuttle twists and turns on its descent.

The deceleration period lasts about 15 minutes.

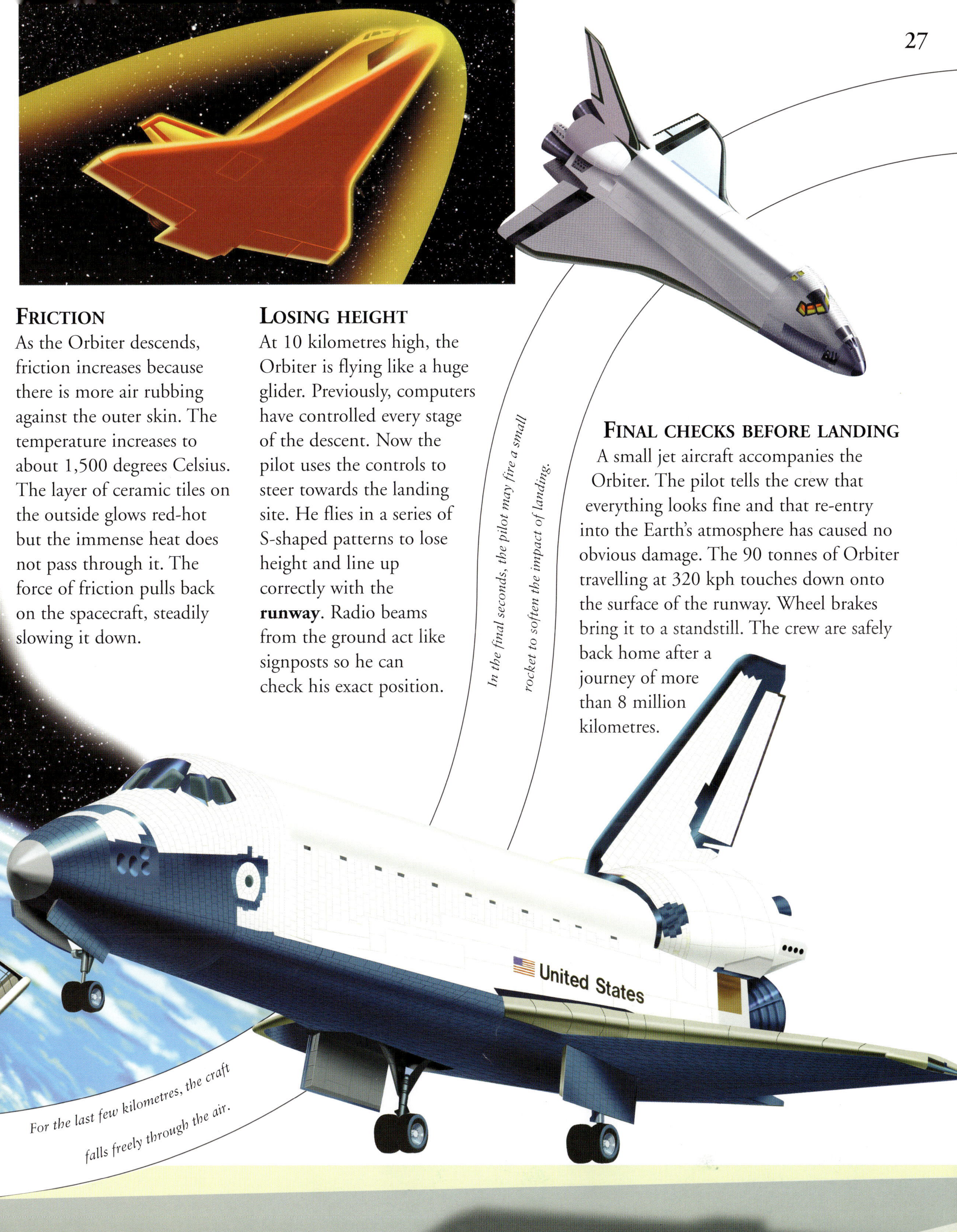

Friction

As the Orbiter descends, friction increases because there is more air rubbing against the outer skin. The temperature increases to about 1,500 degrees Celsius. The layer of ceramic tiles on the outside glows red-hot but the immense heat does not pass through it. The force of friction pulls back on the spacecraft, steadily slowing it down.

Losing height

At 10 kilometres high, the Orbiter is flying like a huge glider. Previously, computers have controlled every stage of the descent. Now the pilot uses the controls to steer towards the landing site. He flies in a series of S-shaped patterns to lose height and line up correctly with the **runway**. Radio beams from the ground act like signposts so he can check his exact position.

In the final seconds, the pilot may fire a small rocket to soften the impact of landing.

Final checks before landing

A small jet aircraft accompanies the Orbiter. The pilot tells the crew that everything looks fine and that re-entry into the Earth's atmosphere has caused no obvious damage. The 90 tonnes of Orbiter travelling at 320 kph touches down onto the surface of the runway. Wheel brakes bring it to a standstill. The crew are safely back home after a journey of more than 8 million kilometres.

For the last few kilometres, the craft falls freely through the air.

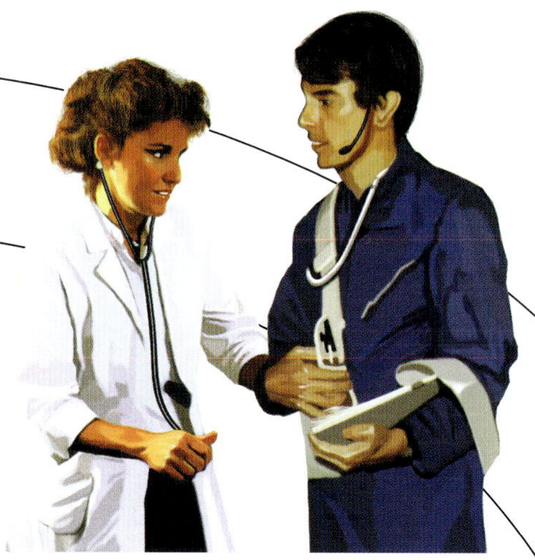

Debrief and Satellite in Action

The Shuttle has returned safely to the ground. Medical staff have checked the crew to make sure they are healthy. In the debriefing room, the crew describe how well the mission went. To help future missions, they compare what actually happened with the plans made before they took off. The solar flare satellite is now in orbit far out from Earth. It will send early warnings about large outbursts of radiation coming from the Sun. Power stations will know to expect problems with the switches that control the flow of electricity. TV companies can alert viewers about interference to their pictures. High-flying jets will cruise at lower altitudes to protect crews and passengers.

Getting back to normal
Doctors at Mission Control are experts in 'space medicine'. They study the effects of staying in space on astronauts' bodies. Returning to Earth means that the astronauts must readjust to gravity again. A particular problem is regaining a sense of balance. Astronauts who stay in space for months take several days to relearn how to walk properly.

The satellite carried 14 million seeds of more than 100 different species of plants.

Scientists expected 'mutations' – damaged seeds growing into weird-shaped plants.

PREDICTING TV INTERFERENCE

"Do not adjust your set: the interference is only temporary". Your TV picture comes from a transmitter less than 40 kilometres away. Extra radiation from the Sun forms electrically-charged layers in the air. TV signals travel further by bouncing off these layers, causing foreign stations to interfere with local TV signals.

READY TO FLY AGAIN

The Orbiter must be ready to fly again within the next nine months. Technicians will check each one of the ceramic tiles that protected it during re-entry. They will also replace many of the internal parts and check the function of every piece of equipment. The booster rockets, recovered from the sea and cleaned, are ready for repacking with fresh solid fuel.

There were far fewer than expected. Seeds are less affected by radiation than humans.

ATMOSPHERIC EFFECTS

Jet fighter planes and the supersonic airliner Concorde fly much higher than ordinary planes. They fly at the upper edge of the atmosphere called the **stratosphere**. The atmosphere shields people on the ground against most radiation from the Sun. During solar flares, these planes must fly lower than usual, inside the protection of the atmosphere. On the ground, electricity cables strung between pylons are affected by solar radiation, upsetting supplies.

The recovered satellite

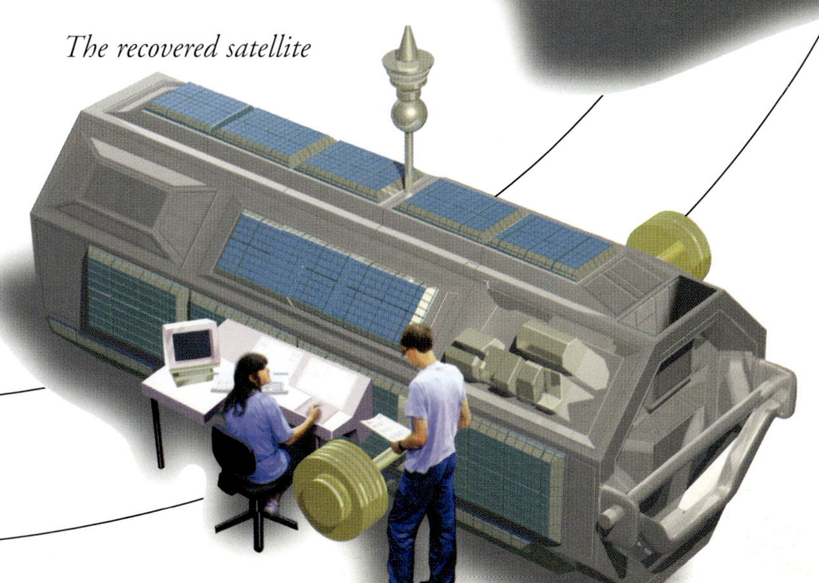

THE INVESTIGATIONS CONTINUE

The results of experiments and tests carried out on Spacelab and on the satellite retrieved from space are checked and analyzed by scientists. The outer surfaces of the two craft are examined for damage from exposure to the Sun and by micrometeorites. The plant seeds that were on the satellite have been distributed to schools around the country. The seeds have now sprouted. Most of the plants are normal, but radiation has caused some of them to develop strange striped leaves.

Glossary

Airlock
A compartment with two doors that allows an astronaut to enter space without the Orbiter losing air.

Antenna
A radio aerial used for receiving or transmitting signals.

Atmosphere
The layer of air surrounding the Earth; breathable air inside the Orbiter.

Attitude thrusters
The small rocket motors that turn the Orbiter around.

Beam
A collection of radio waves or other radiation travelling together in one distinct direction.

Booster rocket
A rocket that helps the main engines at take off.

Cargo bay
The main part of the Orbiter between the crew cabin at the front and the main rocket engines at the rear.

Ceramic
A hard glassy sort of material that can withstand very high temperatures.

Crew cabin
The front part of the Orbiter, consisting of the upper flight deck and the lower crew deck.

Crew deck
The lower part of the Orbiter crew cabin where the crew eat, sleep, and carry out some of the mission experiments.

Electromagnetic radiation
Energy travelling at the speed of light e.g. X-rays, gamma rays, light.

Electrons
Tiny particles charged with electricity and present in all matter.

Extravehicular mobility unit (EMU)
The technical name for a space suit.

Flight deck
The upper part of the Orbiter crew cabin. It contains most of the controls needed to fly the Shuttle and carry out the mission tasks.

Friction
A force that acts against the movement of things that slide against each other.

Fuel cell
A machine that uses oxygen and hydrogen to make electricity and water.

g-force
A force that acts on something that is accelerating. Acceleration at take off is 4g, which is four times normal gravity; this effect increases the weight of things by four times.

Gamma rays
A type of electromagnetic radiation. Gamma rays can penetrate more than 1 metre of solid concrete.

Gyroscope
A heavy spinning wheel. A gyroscope spinning inside a satellite keeps its motion steady and stops it from tumbling.

Hydrogen
The fuel for the main Shuttle engines. It is stored as a liquid so that it takes up the smallest amount of space.

Ions
Tiny particles that have lost electrons. Ions shoot out from solar flares and can travel to the Earth and beyond.

Launch tower
A large tower attached to the Shuttle before launch. It contains a lift for the crew and technicians, together with power and fuel lines.

Living cells
The smallest parts from which all living things are made. Radiation can damage cells so that they do not grow properly.

Magnetic field
The space around a magnet in which the effects of the magnet are felt.

Micro meteoroids
Particles smaller than 2 millimetres in diameter that travel through space at very high speeds.

Microprocessor
A type of computer designed to carry out special tasks.

Mission Control
The place on Earth that stays in constant contact with the crew during the mission.

Mission commander
The crew member in overall charge of the mission.

Mission specialist
The crew member in charge of all the activities and experiments carried out during the mission.

Orbital Manoeuvring System (OMS)
Two rocket engines at the rear of the Orbiter next to the three main engines. They provide the extra thrust needed to enter orbit after take off or to change orbit.

Orbital velocity
28,000 kilometres per hour: the speed a rocket must achieve to leave the Earth and orbit around it.

Orbiter
The winged part of the Shuttle: at the front is the crew deck, followed by the cargo bay, with the rocket engines at the rear.

Oxygen
The gas that is needed for breathing and for fuels to burn. It is stored as a liquid so that it takes up the smallest possible space.

Pilot
The crew member responsible for controlling the flight of the Orbiter, especially during re-entry and landing.

Planets
The nine main bodies orbiting around the Sun: Mercury, Venus, Earth, Mars, Jupiter, Saturn, Neptune, Uranus, and Pluto.

Protons
Tiny particles charged with electricity and present in all matter.

Radar
Finding out where distant objects are by bouncing radio waves from them and analysing the echoes with a microprocessor.

Radar screen
A TV screen that shows the positions of distant objects located by radar.

Radiation
Electromagnetic radiation and beams of tiny particles.

Re-entry
Leaving orbit and descending into the Earth's atmosphere.

Reflector dish
A bowl-shaped piece of metal attached to a radio aerial that focuses the radio waves into a narrow beam.

Remote manipulator system (RMS)
A jointed arm attachment used for pulling satellites into the cargo bay.

Runway
A long flat strip of ground on which the Orbiter lands.

Satellite
Artificial satellites are man-made objects which orbit around a planet or moon: natural satellites are moons which orbit around planets.

Sensor
The part of a satellite which is sensitive to radiation, forces, or particles. Information from sensors show the crew what is happening inside and outside of the Orbiter.

Solar cell
A device which changes the energy from sunlight directly into electricity.

Solar flare
A curved pillar of white-hot gas arcing upwards from the Sun's surface and sending showers of particles and powerful radiation into space.

Solar wind
A stream of particles flowing out into space from the Sun.

Space sickness
A feeling of queasiness and nausea caused by weightless conditions in orbit.

Spacelab
A large container in the Orbiter cargo bay: crew members carry out experiments inside spacelab when there is not enough room inside the crew cabin.

Stratosphere
The upper part of the Earth's atmosphere, between 25 and 50 kilometres above the ground. There is not enough air in the stratosphere to breathe.

Teflon
A type of plastic material which is a very good heat insulator. It was developed for use in space but is now used to coat non-stick frying pans used in everyday cooking.

Thrust
A measure how hard a rocket pushes: 1 tonne of thrust upwards is the same the downward force of gravity on a 1-tonne mass.

Tiles
The curved sheets of ceramic material stuck to the outside of the Orbiter to protect it against heat during re-entry.

Vehicle Assembly Building
The building where the Orbiter, booster rockets, and main tank are fitted together.

Weightless
When masses do not fall down when released: due to zero gravity inside spacecraft.

X-rays
A type of electromagnetic radiation, rather like very powerful radio waves: X-rays can pass through solids and cause damage to living cells.

Zero gravity
When masses are weightless, due to gravity inside a spacecraft being zero.

INDEX